A Compendium of Systems Thinking 2.0

-- The Structure-Behavior Coalescence Approach --

William S. Chao

Structure-Behavior Coalescence

$$\text{Systems Architecture} \equiv \text{Systems Structure} + \text{Systems Behavior}$$

4

CONTENTS

6

ABOUT THE AUTHOR

Dr. William S. Chao is the CEO & founder of SBC Architecture International®. SBC (Structure-Behavior Coalescence) architecture is a systems architecture which demands the integration of systems structure and systems behavior of a system. SBC architecture applies to hardware architecture, software architecture, enterprise architecture, knowledge architecture and thinking architecture. The core theme of SBC architecture is: "Architecture = Structure + Behavior."

William S. Chao received his bachelor degree (1976) in telecommunication engineering and master degree (1981) in information engineering, both from the National Chiao-Tung University, Taiwan. From 1976 till 1983, he worked as an engineer at Chung-Hwa Telecommunication Company, Taiwan.

William S. Chao received his master degree (1985) in information science and Ph.D. degree (1988) in information science, both from the University of Alabama at Birmingham, USA. From 1988 till 1991, he worked as a computer scientist at GE Research and Development Center, Schenectady, New York, USA.

Dr. William S. Chao has been teaching at National Sun Yat-

Sen University, Taiwan since 1992 and now serves as the president of Association of Enterprise Architects, Taiwan Chapter. His research covers: systems architecture, hardware architecture, software architecture, enterprise architecture, knowledge architecture and thinking architecture.

PART I: SYSTEMS THINKING 1.0

The Need of Systems Definition

The need for defining a system arises because any real-life system is inherently complicated. It is impossible to fully comprehend the intricate interrelationships of any system of the real world with its environment, or to describe all its components and each of its details.

Systems definition is an "artifact" created by humans to describe what a system is.

Without a systems definition, everybody has his own saying about a system and never be able to reach a consensus.

Systems Thinking 1.0 Defining a System

Systems thinking 1.0 defines a system, hopefully to be an integrated whole, embodied in its assembled components, their interrelationships with each other and the environment.

As a first example, systems thinking 1.0 defines a *Tree*, hopefully to be an integrated whole, embodied in its assembled components of *Leaves* and *Trunk*, their interrelationships with each other and the environment.

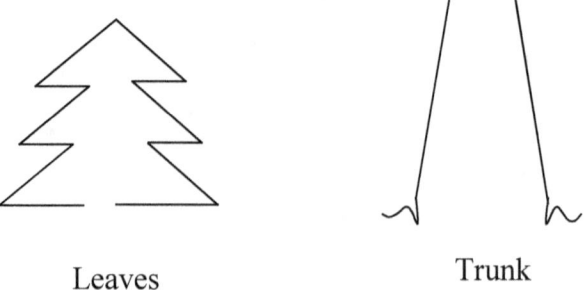

Leaves   Trunk

As a second example, systems thinking 1.0 defines the *Pig*, hopefully to be an integrated whole, embodied in its assembled components of *Head*, *Body* and *Feet*, their interrelationships with each other and the environment.

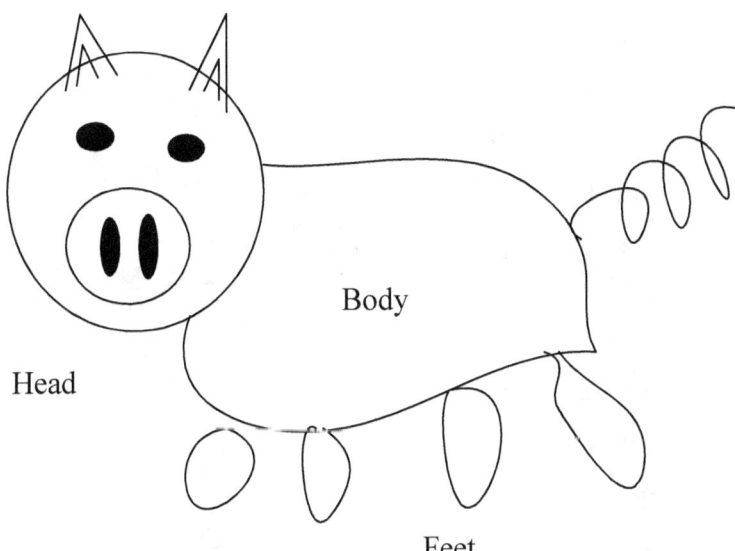

As a third example, some countries may use systems thinking 1.0 to define a *Marriage*, hopefully to be an integrated whole, embodied in its assembled components of *Man* and *Woman*, their interrelationships with each other and the environment.

As a fourth example, other countries may use systems thinking 1.0 to define a *Marriage*, hopefully to be an integrated whole, embodied in its assembled components of *Two Persons Without Distinction As To Their Sex*, their interrelationships with each other and the environment.

As a fifth example, systems thinking 1.0 defines the *Classroom_4069*, hopefully to be an integrated whole, embodied in its assembled components of *Desk_1*, *Chair_1* and *Chair_2*, their interrelationships with each other and the environment.

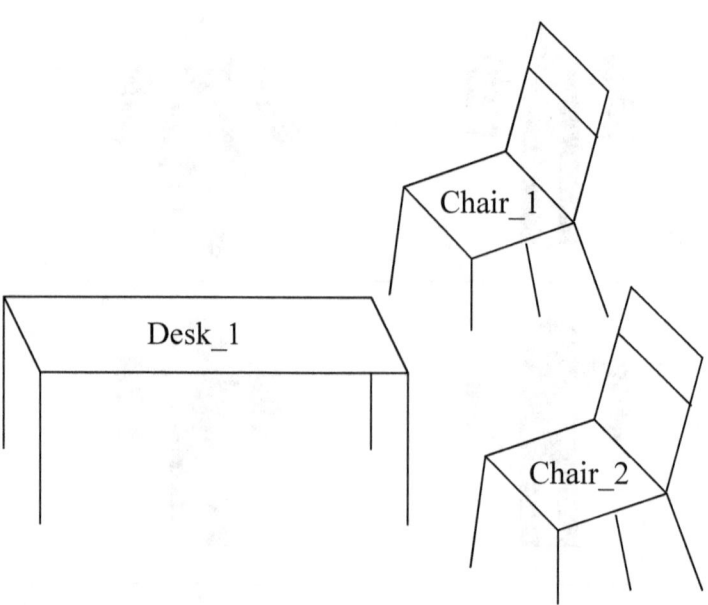

As a sixth example, systems thinking 1.0 defines a *Car*, hopefully to be an integrated whole, embodied in its assembled components of *Body* and *Wheels*, their interrelationships with each other and the environment.

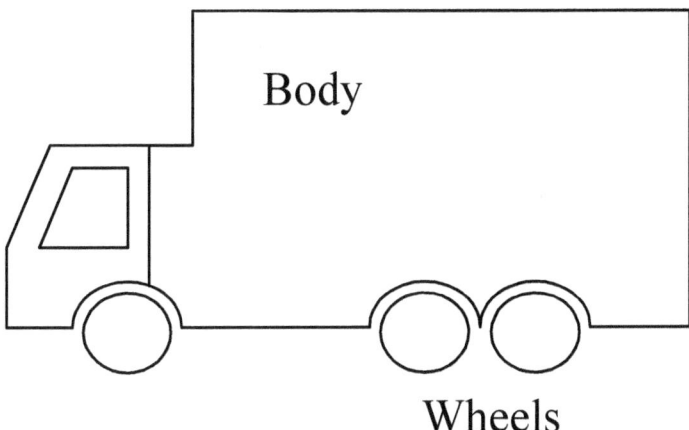

As a seventh example, systems thinking 1.0 defines a *Sandwich*, hopefully to be an integrated whole, embodied in its assembled components of *Fillings* and *Slice_Of_Bread*, their interrelationships with each other and the environment.

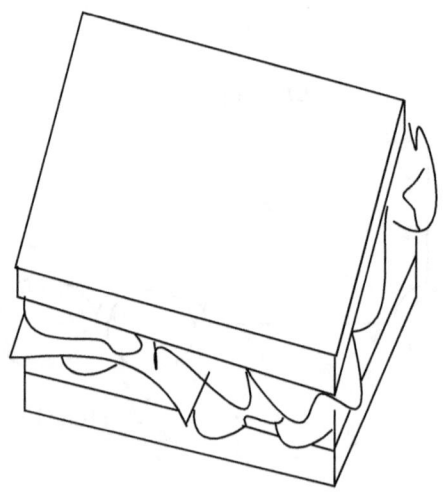

Physical and Virtual Systems

A physical system exists in the physical world. A physical system is also called a concrete or real system. For example, a *Telephone* composed of *Microphone*, *Earphone* and *Keypad* is a physical, concrete, or real system.

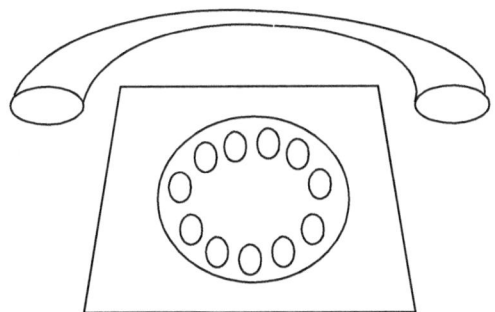

A virtual system is a system that is composed of non-physical components, i.e., ideas, thoughts, or notions. A virtual system exists in the virtual, abstract, or notional world. For example, the *"Snow White and the Seven Dwarfs"* fairy tale composed of *"Snow White"* and *"Seven Dwarfs"* is a virtual, abstract, or notional system.

Boundary and Environment of a System

We scope a system by defining its boundary. All components of the system are inside the boundary while the environment is outside the boundary.

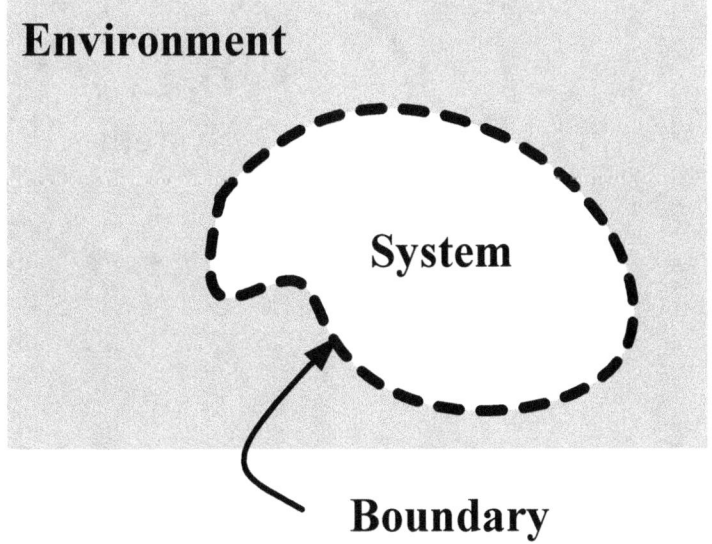

The environment is also known as the surroundings. A system may or may not interrelate with the environment. An open system interrelates with the environment through the exchange of matter, energy, data, information, or message.

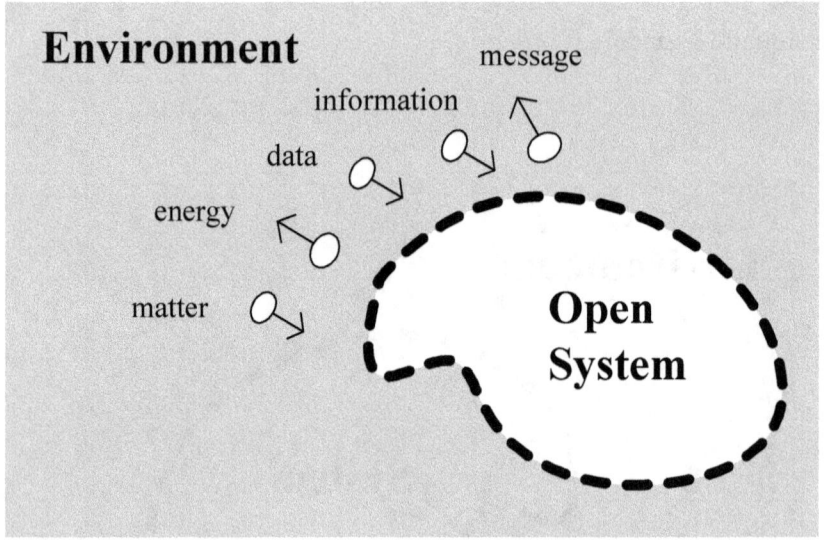

An isolated system does not interrelate with the environment at all. There is no exchange of matter, energy, data, information, or message between the isolated system and the environment.

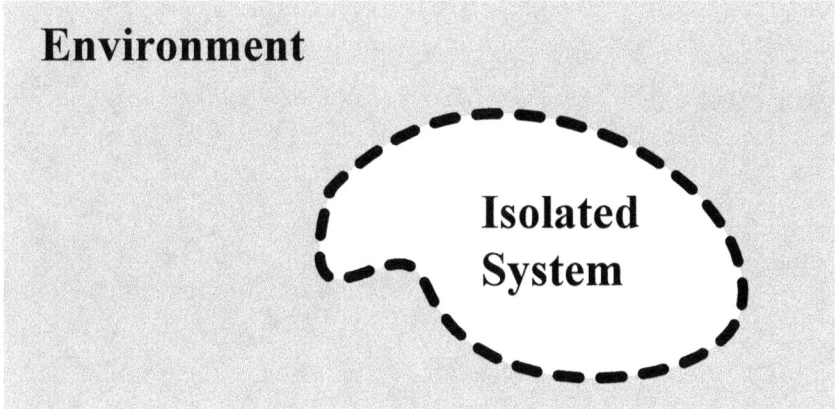

PART II: SHORTCOMINGS OF SYSTEMS THINKING 1.0

Systems Thinking 1.0 Does Not Describe the Integration of Systems Structure and Systems Behavior

Systems structure and systems behavior are the two most significant views of a system. In order to achieve a truly integrated whole of a system, we first need to integrate the systems structure and systems behavior together.

In other words, integration of systems structure and systems behavior results in the integration of a whole system.

Since systems thinking 1.0 does not describe the integration of systems structure and systems behavior, very likely it only hopes and will never be able to really form an integrated whole of a system.

Systems Thinking 1.0 is Powerless in Defining a System Appropriately.

Systems thinking 1.0 does not describe the integration of systems structure and systems behavior, very likely it only hopes and will never be able to really form an integrated whole of a system.

In this situation, systems thinking 1.0 is powerless in defining a system appropriately.

PART III: THE STRUCTURE-BEHAVIOR COALESCENCE APPROACH

Structure-Behavior Coalescence Means to Integrate the Systems Structure and Systems Behavior

Systems structure and systems behavior are the two most prominent views of a system, integrating the systems structure and systems behavior is apparently the best way to achieve an integrated whole of a system.

If we are not able to integrate the systems structure and systems behavior, then there is no way that we are able to integrate the whole system.

Structure-behavior coalescence (SBC) provides an elegant way to integrate the systems structure and systems behavior of a system. In other words, SBC facilitates an integrated whole of a system.

Core Theme of Structure-Behavior Coalescence

The core theme of structure-behavior coalescence is: "Systems Architecture = Systems Structure + Systems Behavior."

Systems Structure X $+$ Systems Behavior X

One systems structure will draw forth one systems behavior. That is, the systems behavior is attached to or built on the systems structure in the SBC approach.

In other words, the systems behavior can not exist alone; it must be loaded on the systems structure just like a cargo is loaded on a ship.

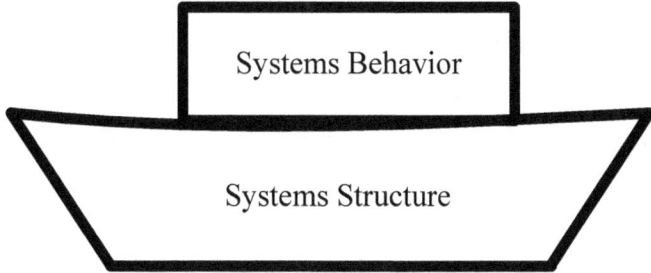

SBC Architecture Description Language

SBC architecture description language (SBC-ADL) uses six fundamental diagrams to describe the integration of systems structure and systems behavior of a system. These diagrams are: a) architecture hierarchy diagram (AHD), b) framework diagram (FD), c) component operation diagram (COD), d) component connection diagram (CCD), e) structure-behavior coalescence diagram (SBCD) and f) interaction flow diagram (IFD).

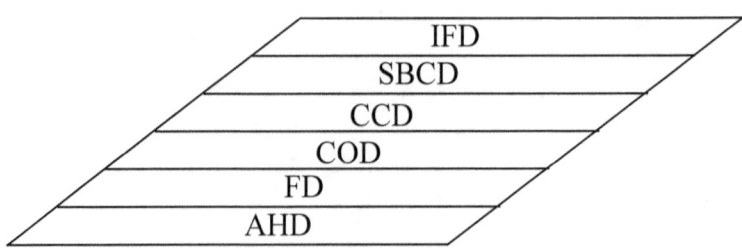

AHD, FD, COD and CCD to Draw Forth the SBCD and IFD

SBC-ADL uses AHD, FD, COD and CCD to depict the systems structure of a system.

SBC-ADL uses SBCD and IFD to depict the systems behavior of a system.

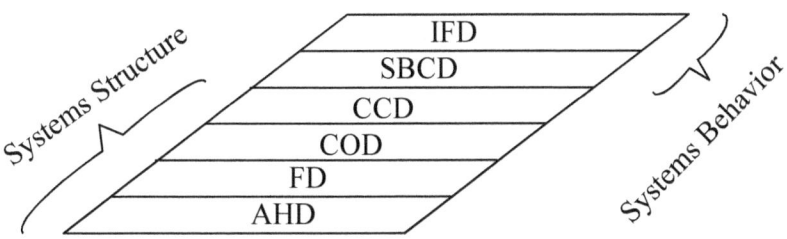

Undoubtedly, AHD, FD, COD and CCD draw forth the SBCD and IFD. Therefore, SBC architecture description language gracefully describes the integration of systems structure and systems behavior of a system.

SBC-ADL represents a knowledge repository of a system. Stakeholders can submit and acquire knowledge to and from this knowledge repository.

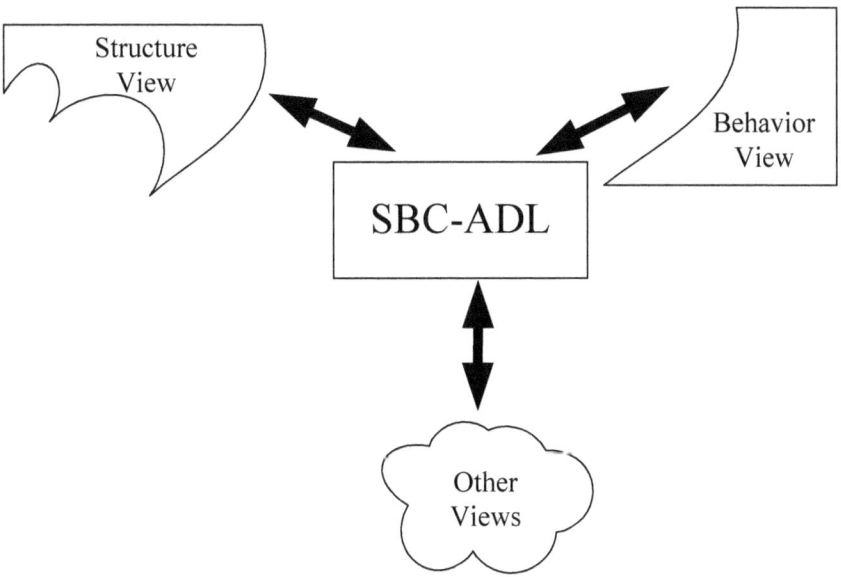

Interaction

An interaction represents an indivisible and instantaneous handshake or communication between two agents. The caller agent (either external environment's actor or component) communicates with the callee agent (component) through the interaction.

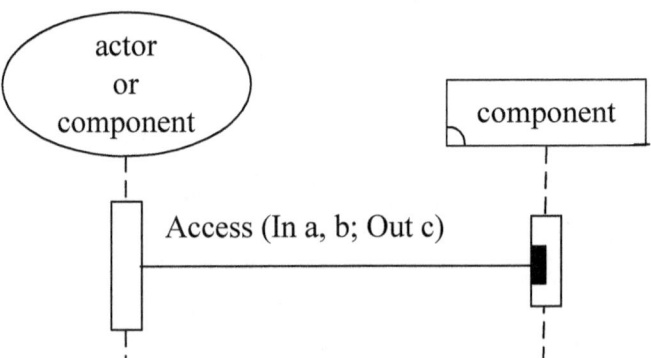

There are two ports, i.e., calling port or called port, of an interaction. The caller agent owns the "calling port" of the interaction. The callee agent owns the "called port" of the interaction.

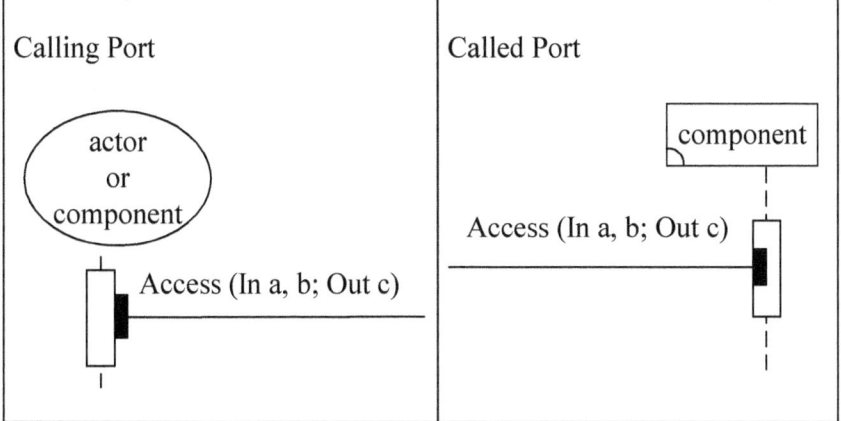

Interactions among Components and Actors to Draw Forth the Systems Behavior

In a system, if the components, and among them and the external environment's actors to interact (or handshake), these interactions will draw forth the systems behavior.

We conclude that "interaction" plays an important factor in integrating the systems structure and systems behavior for a system.

The overall behavior of a system consists of many individual behaviors. Each individual behavior represents an execution path. We use an interaction flow diagram (IFD) to demonstrate this individual behavior.

Collection of All Interaction Flow Diagrams Defines the Systems Architecture

The collection of all interaction flow diagrams defines the integration of systems structure and systems behavior of a system.

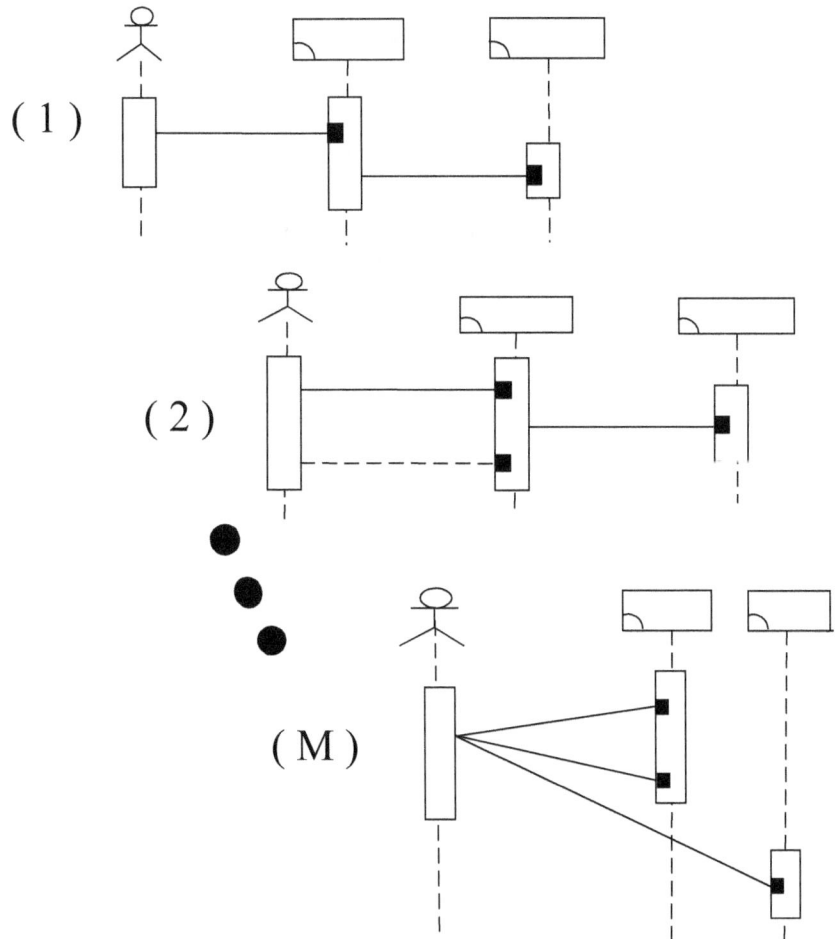

That is, the collection of all interaction flow diagrams defines the systems architecture.

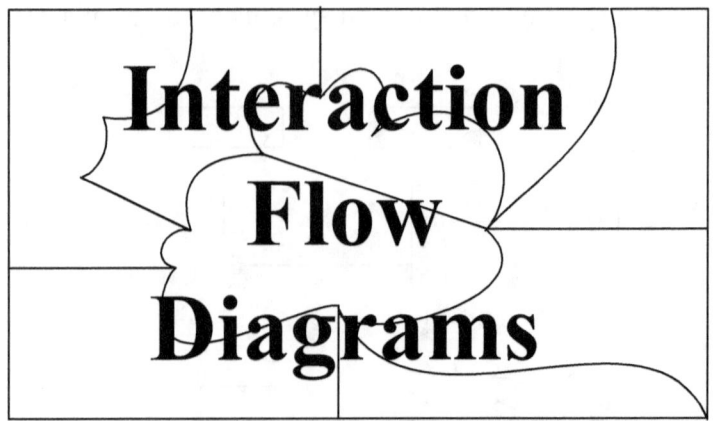

Systems architecture (SA) represents a knowledge repository of a system. Stakeholders can submit and acquire knowledge to and from this knowledge repository.

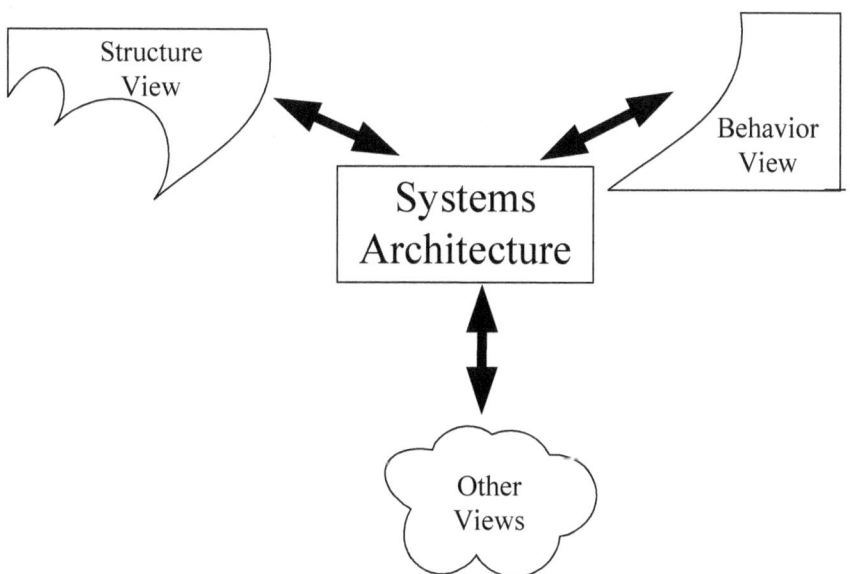

So, the collection of all interaction flow diagrams represents a knowledge repository of a system. Stakeholders can submit and acquire knowledge to and from this knowledge repository.

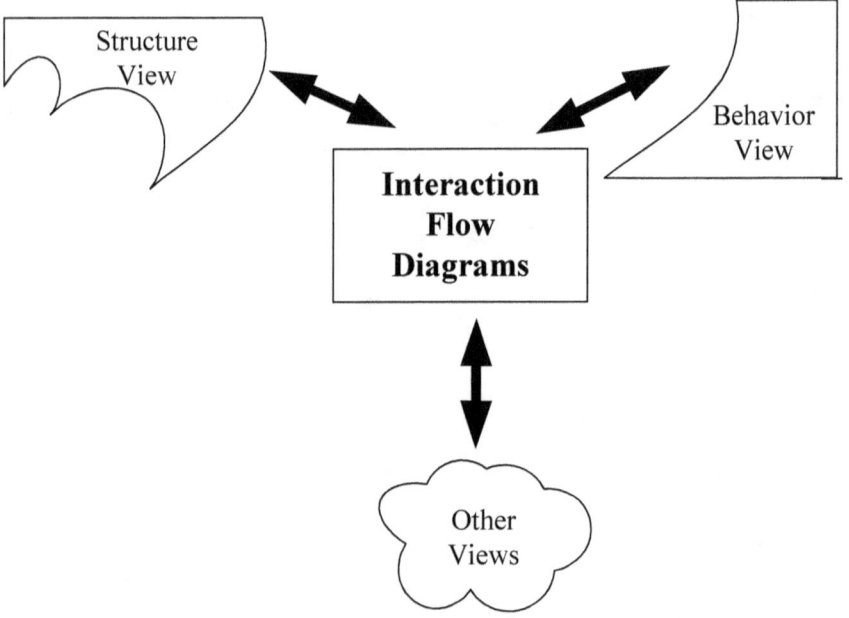

PART IV: SYSTEMS THINKING 2.0

Systems Thinking 2.0 Defining a System

Systems thinking 2.0 defines a system, through the SBC architecture description language, truly to be an integrated whole, embodied in its assembled components, their interactions with each other and the environment.

A system defined by systems thinking 2.0 has the following characteristics: 1) it emphasizes the system's structure-behavior coalescence; 2) it is a truly integrated whole; 3) it is embodied in its assembled components; 4) components are interacting with each other and the environment; and 5) it uses structural decomposition rather than functional decomposition.

Systems Thinking 2.0 Is Architectural Thinking

Since systems thinking 2.0 uses the SBC architecture description language to define a system, systems thinking 2.0 is also called architectural thinking.

Systems Thinking 2.0
Architectural Thinking

PART V: SYSTEM THINKING 2.0 ADVOCATES MATERIALISM-IDEALISM UNITY

Materialism

Materialism is a form of philosophical monism which holds that only the matter or physical is real, and that the mental or spiritual can be reduced to the physical. Materialism is closely related to physicalism.

唯物論

Idealism

Idealism is the philosophical view that a variety of existing things can be explained in terms of mind or spirit.

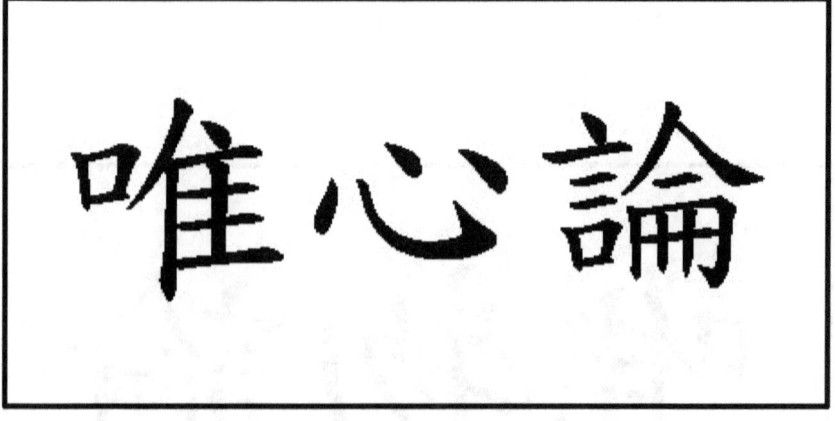

Idealism is also called spiritualism or phenomenalism or mentalistic monism.

Materialism-Idealism Unity

Materialism-idealism unity means to integrate the material and mind.

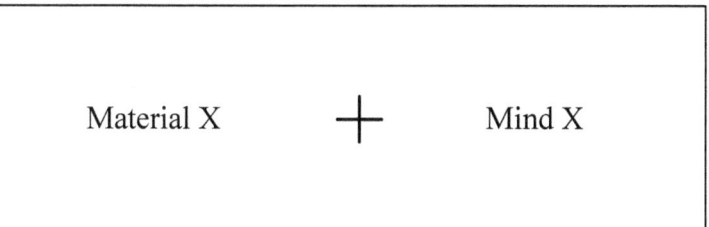

System thinking 2.0 (Architectural thinking) advocates materialism-idealism unity.

心物合一

PART VI: SYSTEM THINKING 2.0 ADVOCATES SUBSTANCE-FUNCTION UNITY

Substance

In Chinese philosophy, substance or original reality is the cause of all transformations in this universe.

Function

In Chinese philosophy, function is the myriad of manifestations of substance in this universe.

Substance-Function Unity

The idea that substance and function are in fact one unit, is a metaphysical claim that is key to Confucianism, Zen and Taoism.

Substance-Function unity means to integrate the substance and function.

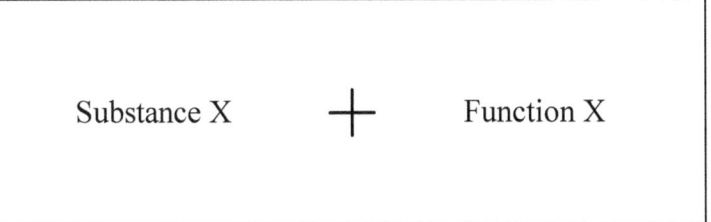

System thinking 2.0 (Architectural thinking) advocates substance-function unity.

體用不二

PART VII: SYSTEM THINKING 2.0 ADVOCATES STRUCTURALISM-FUNCTIONLISM UNITY

Structuralism

In sociology, anthropology and linguistics, structuralism is the theory that elements of human culture must be understood in terms of their relationship to a holistic system or structure. It works to uncover the structures that underlie all the things that humans do, think, perceive and feel.

結構主義

Functionalism

Functionalism, is a framework for building theory that sees society as a complex system whose parts work together to promote solidarity and stability. Functionalism addresses society as a whole in terms of the function. In the most basic terms, it simply emphasizes "the effort to impute, as rigorously as possible, to each feature, custom, or practice, its effect on the functioning of a supposedly stable, cohesive system".

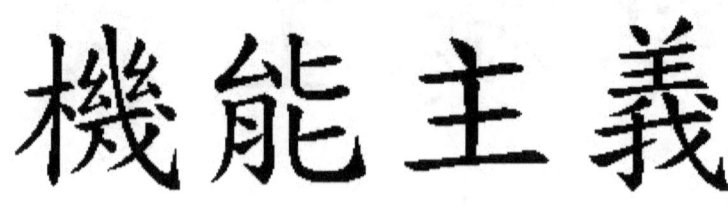

Structuralism-Functionalism Unity

Structuralism-functionalism unity means to integrate the structure and function.

Structure X $+$ Function X

System thinking 2.0 (Architectural thinking) advocates structuralism-functionalism unity.

結構機能合一

www.ingramcontent.com/pod-product-compliance
Lightning Source LLC
Chambersburg PA
CBHW070939180526
45168CB00003B/1100